Let's Discover Our Seashores, Singapore!

Exploring the amazing creatures found on our seashores, with one of Singapore's foremost marine biologists!

Authors: Chou Loke Ming and Diana Chou
Illustrator: Eliz Ong

WS Education

NEW JERSEY · LONDON · SINGAPORE · BEIJING · SHANGHAI · HONG KONG · TAIPEI · CHENNAI · TOKYO

Published by

WS Education, an imprint of
World Scientific Publishing Co. Pte. Ltd.
5 Toh Tuck Link, Singapore 596224
USA office: 27 Warren Street, Suite 401-402, Hackensack, NJ 07601
UK office: 57 Shelton Street, Covent Garden, London WC2H 9HE

National Library Board, Singapore Cataloguing in Publication Data
Name(s): Chou, L. M.   Chou, Diana, author.   Ong, Eliz, illustrator.
Title: Let's discover our seashores, Singapore! : exploring the amazing
    creatures found on our seashores, with one of Singapore's foremost
    marine biologists! / authors, Chou Loke Ming and Diana Chou ;
    illustrator, Eliz Ong.
Other Title(s): Exploring the amazing creatures found on our seashores,
    with one of Singapore's foremost marine biologists!
Description: Singapore : WS Education, 2022
Identifier(s): ISBN 978-981-12-5102-3 (hardcover)
    978-981-12-5113-9 (paperback)
    978-981-12-5103-0 (ebook for institutions)
    978-981-12-5104-7 (ebook for individuals)
Subject(s): LCSH: Seashore biology--Singapore--Juvenile literature.
    Seashore ecology--Singapore--Juvenile literature.
Classification: DDC 578.7699095957--dc23

British Library Cataloguing-in-Publication Data
A catalogue record for this book is available from the British Library.

LET'S DISCOVER OUR SEASHORES, SINGAPORE!
**Exploring the Amazing Creatures Found on Our Seashores,
with One of Singapore's Foremost Marine Biologists!**

Copyright © 2022 by Chou Loke Ming & Diana Chou
All rights reserved.

ISBN 978-981-125-102-3 (hardcover)
ISBN 978-981-125-113-9 (paperback)
ISBN 978-981-125-103-0 (ebook for institutions)
ISBN 978-981-125-104-7 (ebook for individuals)

For any available supplementary material, please visit
https://www.worldscientific.com/worldscibooks/10.1142/12689#t=suppl

Desk Editor: Amanda Yun

Printed in Singapore

The ready support that we received from colleagues and friends
in the preparation of this book gave us immense encouragement and confidence.
We thank the following for reviewing the content and providing valuable feedback:
Dr. Beverly Goh
Dr. Huang Danwei
Dr. Lionel Ng
Dr. Tan Lik Tong
Dr. Karenne Tun
Dr. Yap NWL

This book relied heavily on pictures of the plants and animals seen on the beach,
and we thank the following for generously allowing us to use their pictures:
Ms. Cheo Pei Rong
Mr. Steven Lim Keng Hwee
Mr. Jonathan Tan

Those who helped in one way or another are:
Benjamin Eng Jun En
Julian Eng Yi En

Picture credits:
Cheo Pei Rong p9 11, 14, 16 17, 36, 39, 48 49, 51 52, 56, 60, 66 67.
Steven Lim Leng Hwee p8, 18 35, 37, 39, 44 47, 54 55, 57 58, 60, 62 65, 67 71.
Jonathan Tan p14 15, 36, 38, 40 43, 48, 50 51, 53, 57, 60, 67.

Introduction

Hello! We are Prof. Chou, a marine biologist, and Diana Chou, a music and pre-school educator.

Did you know that Singapore has an abundance of interesting marine life right at our doorstep? When the tide is low, you can see them and be amazed by how they live and move. Our seashores have a large variety of marine creatures, all of which are interesting. The more observant you are, the more you'll get to see. Just remember that you are walking in their territory and take care not to accidentally step on them or take them home. When you see a creature, spend some time watching what it is doing. Each creature has an important role in the seashore habitat because they interact with one another and leaving them as they are means that others will get to see them too. In this book, you'll learn more about 25 of the creatures that live along Singapore's shoreline.

A Guide to Experiencing the Bonus Features in This Book

See a QR code on a page? Scan it to access the page's bonus feature!

What to do when you meet wildlife at the beach

When you visit the beach at low tide,
By nature's rules you should abide!
Always treat them with respect.
Learn through oberservation and watch where you trek!

Take only pictures, not creatures from their homes
And leave brightly-coloured animals well alone.

Wear closed shoes to avoid cuts, bites or stings.
And always be alert towards your surroundings.
Stay hydrated when the sun is up,
And when you see trash — pick it up!
Do help to keep our beaches clean
But be aware of the tide as it will come in.

✓ DOs

1. Walk slowly so that you don't alarm the sea creatures.
2. Stop to observe what they are doing and how they live.
3. Keep a suitable distance between you and the creature.
4. Take pictures of the creatures, especially those that you are not familiar with.
5. Help to remove trash, discarded nets, and fishing gear.
6. Get help from an adult to remove sharp objects like fishing hooks.

✗ DON'Ts

1. Do not pick up the creatures. This will only stress them. Also some can sting or bite.
2. Avoid stepping on any creature.
3. Do not move any creature. Changing its position interrupts its intended activity.
4. Do not dig into the sand or mud to look for creatures. Let the creatures emerge on their own accord.
5. Never collect creatures and put them in a pail for viewing later, because they may kill or eat each other, and get weak when released higher up the shore.

Got that? Now let's take a look at some of the creatures that live on Singapore's Beaches!

CONTENTS

Marine Life in this Book

Sea Lettuce	*Ulva sp.*	8
Fern Seagrass	*Halophila spinulosa*	10
Haddon's Carpet Anemone	*Stichodactyla haddoni*	14
Swimming Anemone	*Boloceroides mcmurrichi*	16
Tube Anemone	*Pachycerianthus sp.*	18
Flowery Soft Coral	*Dendronephthya sp.*	20
Spiky Sea Pen	*Pteroeides sp.*	24
Tube Worm		26
Scale Worm		28
Coastal Horseshoe Crab	*Tachypleus gigas*	30
Orange-Striped Hermit Crab	*Clibanarius infraspinatus*	32
Ball Moon Snail	*Neverita didyma*	34
Pearl Conch	*Strombus turturella*	36
Noble Volute	*Cymbiola nobilis*	38

Starry-mouthed Nudibranch	*Bornella stellifer*	40
Eight Arm Seastar	*Luidia maculata*	44
Biscuit Sea Star	*Goniodiscaster scaber*	48
Painted Sand Star	*Astropecten vappa*	50
Knobbly Sea Star	*Protoreaster nodosus*	52
Cake Sand Dollar	*Arachnoides placenta*	54
Thorny Sea Cucumber	*Colochirus quadrangularis*	56
Pink Warty Sea Cucumber	*Cercodemas anceps*	58
Smooth Sea Cucumber	*Acaudina sp.*	62
White Salmacis Urchin	*Salmacis sphaeroides*	66
Acorn Worm		68

Sea Lettuce

Scientific name: *Ulva* sp.

Plant or Animal? Or?

Seaweeds are definitely not animals
But they're not classified as plants either.
They are instead considered as **algae**
Due to their different function and structure.

Seaweeds are grouped according to colour —
Mainly brown, red and green.
Sea Lettuce is a common green seaweed.
At low tide, it creates a green carpet scene!

What's in the scientific name?

"sp." is the abbreviation for "species", which means "types". It is used when the exact species is not known.

A mixed mass of green and red seaweed.

Let's Identify!

Sea Lettuce can grow in thick clumps,
Its structure is translucent, soft and thin.
When the tide is out, they flop on the sand,
But become suspended when the water comes in.

Sea Lettuce flopping on the sand like a green carpet when the tide is out.

Fish Bites

Seaweeds, like plants, are capable of photosynthesis. However, they are classified as algae as they do not have roots, stems, leaves or flowers.

Sea Lettuce or *Ulva*, is recognised by its thin, translucent structure.

Fern Seagrass

Scientific name: *Halophila spinulosa*

Fern Seagrass, characterised by multiple paired leaves along a thin stem, resemble a fern. Fern Seagrass can be seen here growing next to a patch of red seaweed.

Plant or Animal?

Plant or animal, what do you think?
Seagrasses photosynthesize energy from the sun,
have roots, stems and leaves,
And are adapted to seawater —
It is a plant!

Red seaweed

What is Photosynthesis?
Photosynthesis refers to the process where plants use energy from the sun to turn carbon dioxide and water into oxygen and carbohydrates (food).

Fern Seagrass

Let's Identify!

Fern Seagrasses have leaves arranged in pairs,
Opposite each other, along a thin stem.
They are primary food producers
And many animals feed on them.

The leaves are quite small —
2 cm long and 0.5 cm wide —
They are common on the seashore
And are best seen at low tide.

Fern Seagrass is not to be confused with the Spoon Seagrass (shown on the left), which can be recognised by its larger, oval, spoon-shaped leaves.

Fish Bites

There are about 12 species of seagrasses in Singapore. This is half the number of species known throughout the Indo-Pacific region.

Fern Seagrass can be found in the South China Sea region and also in northern and western Australia.

Here's a picture of Fern Seagrass and Spoon Seagrass growing alongside each other. Can you tell which is which?

Spoon Seagrass →

← Fern Seagrass

Haddon's Carpet Anemone

Scientific name: *Stichodactyla haddoni*

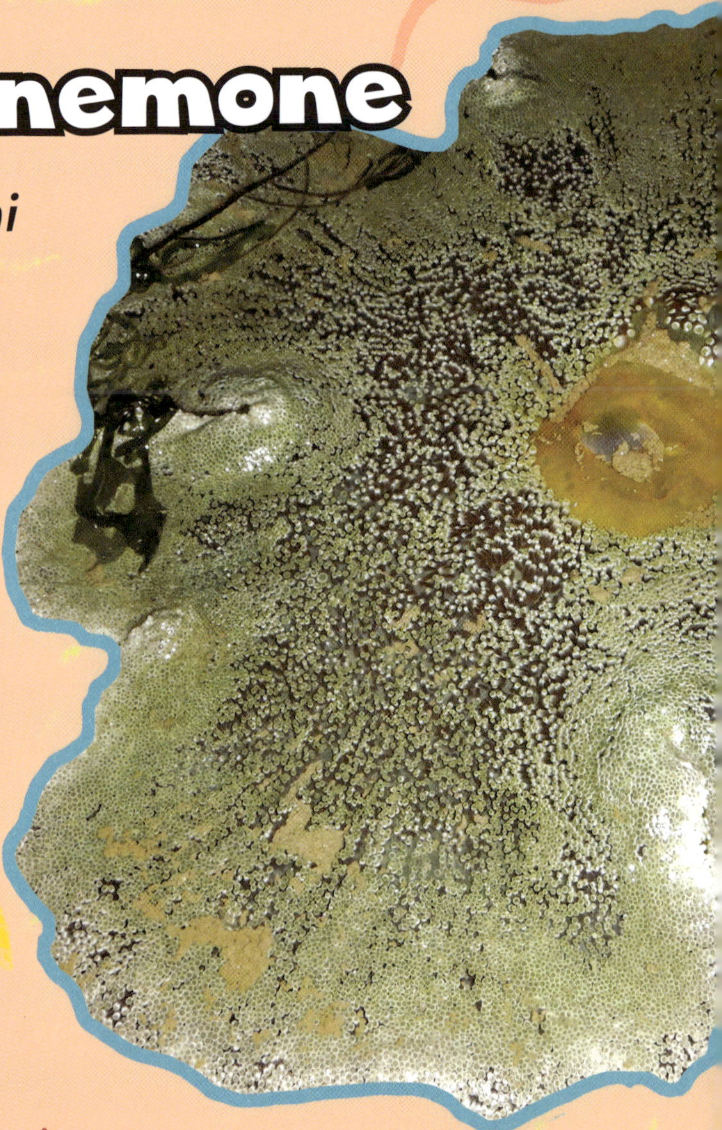

Let's Identify!

My top surface is called the **oral disc**,
And under me is my body column.
I extend it to anchor in the sand
So strong waves don't become a problem.

My disc is covered with short, rounded tentacles,
I look like a carpet spread out on the sand.
I can grow 40—80 cm in diameter
And I can look quite grand.

Here, a Peacock-tail Shrimp is seen moving freely on the anemone's disc.

Here's a Carpet Anemone in the process of swallowing a sand dollar whole!

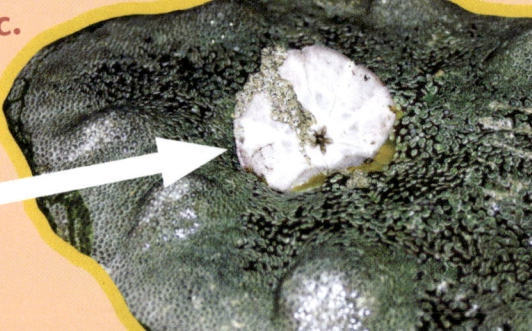

Feeding Time!

My disc can look green or purple,
But commonly it is a shade of grey.
The numerous tentacles on my disc
Help me catch my prey.

My tentacles have stings
That stun and paralyse my prey,
These stings make my tentacles feel sticky
So my prey will not go astray.

They carry my prey to my mouth
Like a ball going into its goal.
Once they reach my mouth,
I will swallow them whole!

Using its tentacles, this anemone pulls a crab carcass into its mouth.

Fish Bites

Nature's Traitors?
Some creatures like the Peacock-tail Shrimp live among the tentacles of the Carpet Anemone and are not affected by its stings.

They help to attract prey to the anemone and in return, get to feed off the scraps pushed out by the anemone.

What's in the name?
The species is named in honour of Alfred C. Haddon, a famous anthropologist who worked on sea anemones from the Great Barrier Reef.

Swimming Anemone

Scientific name:
Boloceroides mcmurrichi

Let's Identify!

My top surface is called the **oral disc**,
And under me is my body column.
I can attach myself to rock or seaweed,
But most of the time, I lie freely on the bottom.

My disc is covered with long pointed tentacles,
That are soft but can look spikey.
My colour is mostly brownish
And some might think I look untidy.

I'm usually 5–8 cm in diameter.
I can swim freely through the water
By flapping my tentacles
In a coordinated rhythmic manner.

Look at these tentacles! The tentacles have a broader base, narrowing to a pointed tip.

Superpowers!

I think I'm quite amazing
As I have a few superpower traits.
Firstly, when a tentacle drops off,
Another **regenerates**!

A dropped tentacle will continue to wriggle,
And this may distract my enemy.
Also with time, a dropped tentacle
Can develop into a new anemone!

Lastly, I can make my own nutrients
With the help of single-celled algae.
They reside in my tissues
And they're called
***zooxanthellae**.

This Swimming Anemone is resting on a seaweed patch.

Fish Bites

The Swimming Anemone's mouth is in the centre of its disc, surrounded by the long tentacles.

....

Photosynthesis
This is the process by which the zooxanthellae provide the Swimming Anemone with nutrients.

*Let's Pronounce zooxanthellae:
"zoo-zen-theh-lae"

Tube Anemone

Scientific name: *Pachycerianthus* sp.

Let's Identify!
I have many long tentacles,
And when the tide is out I droop.
I look like a mess of noodles but
I'm an anemone that lives in a tube.

A Builder by Nature
I build my own tube
With mucus and fibrous threads I secrete,
So that when there is danger
Into it I can retreat.

Fish Bites

The Tube Anemone is a solitary creature. Each tube only has 1 individual. They do not share tubes.

Equipped to Survive
A Tube Anemone's tentacles carry stings. When touched, they discharge a barbed thread with toxin into the victim. This is how they capture prey, and also protect themselves against predators.

The outer tentacles are used more for defence. The inner tentacles help to move food towards its mouth.

*Let's Pronounce Polyps: "Por-Lips"

Flowery Soft Coral

Scientific name: *Dendronephthya* sp.

Plant or Animal?

Corals look like plants but are actually animals.
A colony of individuals called **polyps***
Combine together to form a structure
Like a house constructed from bricks.

Each polyp has a mouth and
Around the mouth many tentacles sit.
Food is taken in through the mouth and
Waste is also expelled through it!

Let's Identify!

We look like a bunch of flowers
With branches that are translucent white.
The tips support many of us polyps
Coloured orange or red — so bright.

We have a thick central stalk
That's attached to a hard surface
So we do not get tossed by waves.
It's our way of staying in place!

Our texture is rubbery and soft,
In water we extend and wave about;
At low tide we slump on the sand —
But we're alive so don't pull us out!

Unable to hold itself up, the Flowery Soft Coral will slump on the sand when the tide is out.

The striking colours of the branch tips make the coral look like a bunch of flowers.

Fish Bites

A Flowery Soft Coral colony can grow to 10–15 cm in size.

....

Corals feed on plankton and tiny food particles in the water.

....

Soft corals do not produce a hard calcium carbonate skeleton, unlike hard corals. When they die, their structure breaks up and disintegrates.

The coral needs the incoming tide to fully expand its branches and resume feeding on plankton.

Spikey Sea Pen

Scientific name: *Pteroeides* sp.

Spikey Sea Pens may get uprooted during low tide — like the one in this picture.

Let's Identify!

We speak as many, not one
As we are many **polyps** in a **colony**.
We belong in the same group as
The coral and anemone.

Each colony has a thick central stalk
That can reach 20 cm in length,
Inside sits a rod of calcium carbonate
Which gives the stalk its strength.

Our upper half is fully exposed
And has leaf-like structures
Arranged on either side of the stalk,
With a look that resembles feathers.

Our lower half is anchored down
Often buried in the sand.
This lets us easily withdraw in
When against predators we defend.

Defence!

Our leaf-like structures have sharp points
And many stings — have you ever wondered why?
This is so that we can stun creatures
That may attack us when they swim nearby.

But some predators like the **nudibranchs**
Can defend against our stinging cells.
Nudibranchs, pronounced as "new-dee-branks",
Are sea slugs without shells.

The lower part of the Spikey Sea Pen anchors it into the sand.

The leaf-like structures make the Spikey Sea Pen resemble a feather quill (quills were the writing 'pens' of the past, and were often made from bird feathers), hence its name. These structures fully extend when the animal is submerged.

Fish Bites

Sea pens, corals and anemones belong to a group called Cnidaria (pronounced as 'nye-day-ria'; the 'c' at the beginning is silent!)

However, the sea pen and most coral structures are made up of a colony of individuals known as polyps, each with a mouth, stomach and tentacles. An anemone is one huge polyp with one mouth, one stomach and many tentacles.

Tube Worm

There are many different species of Tube Worms.

Let's Identify!

We are abundant on the seashore.
At low tide, you see us all around —
Look out for our tube-like structures
That stick up from the ground.

Nature's Builders

Our tubes are like our houses;
We take pride in building them.
Some are papery and some are like thick bark
With diameters up to 1 cm.

If you look at the shore in the photo, you will see tubes sticking out all around. The Tube Worms remain hidden in these protective tubes and are not easy to spot. We should not dig them out as that will injure them and destroy their protection.

The tubes stick out about 10 cm from the ground.

Some of us use mucus to build our houses,
Adding sand or shell into the mix.
Others use calcium carbonate —
Don't we remind you of the 3 Little Pigs?

We stay inside for protection
So that we can remain moist.
Stay and watch the tube opening —
Sometimes, out of the tube we'll hoist.

Fish Bites

Tube Worms feed on a variety of small organisms, and are in turn eaten by larger creatures.

....

The Tube Worm will immediately retract deep down its tube when it senses danger.

The tube opening is at the top.

A Tube Worm hoisting a little bit of its front end out the tube.

Scan this QR code to enjoy a trip to the beach to watch tube worms!

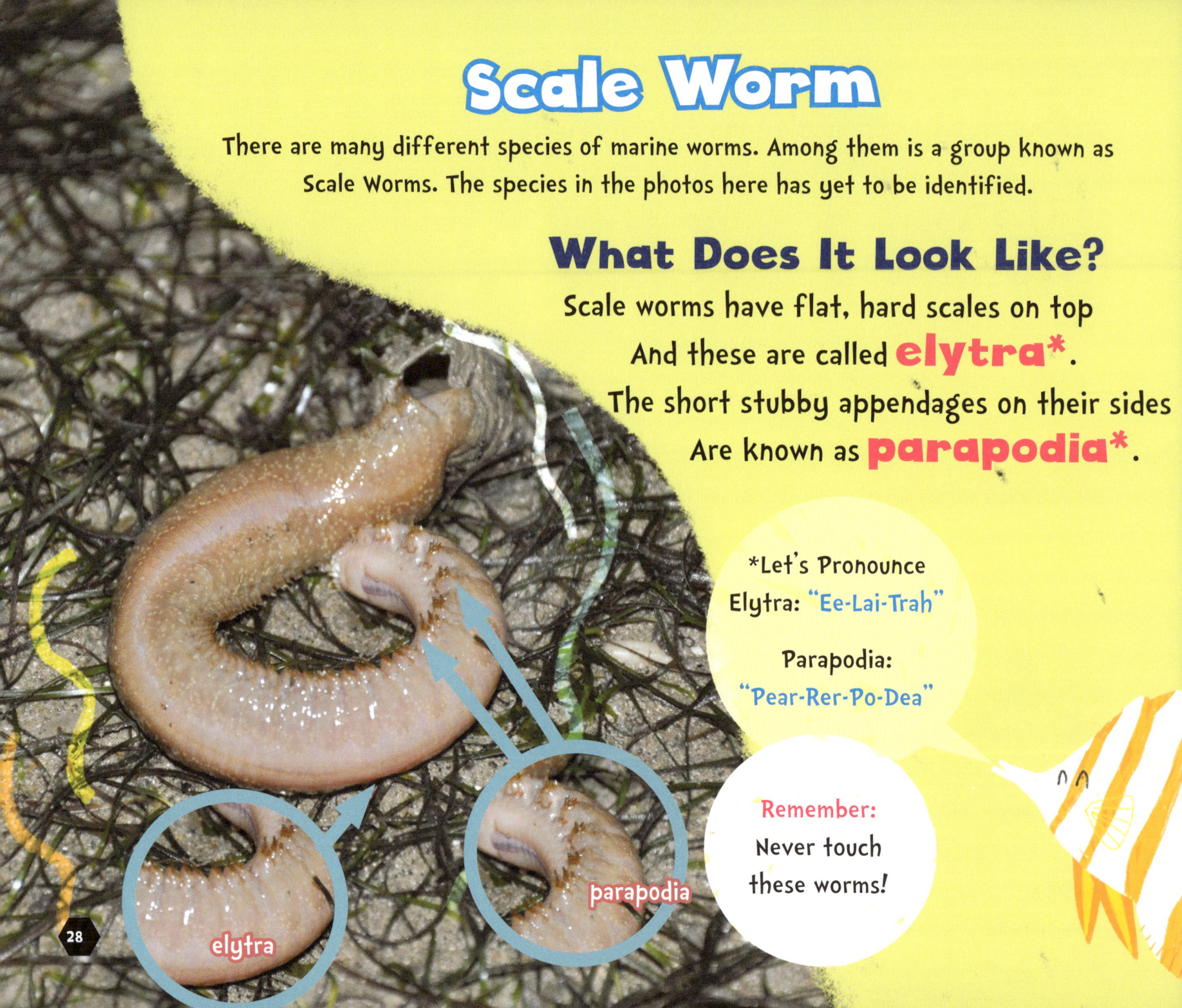

Scale Worm

There are many different species of marine worms. Among them is a group known as Scale Worms. The species in the photos here has yet to be identified.

What Does It Look Like?

Scale worms have flat, hard scales on top
And these are called **elytra***.
The short stubby appendages on their sides
Are known as **parapodia***.

*Let's Pronounce
Elytra: "Ee-Lai-Trah"

Parapodia: "Pear-Rer-Po-Dea"

Remember: Never touch these worms!

elytra

parapodia

Fish Bites

Each parapodia bears numerous bristles that give these worms a spiny look. In general, they are called bristle worms. Scale worms have elytra and are a subset of bristle worms.

Can you see the worm's repetitive body segments, and the bristly looking parapodia on either side of its body?

The shape of the parapodia varies depending on the mode of locomotion of each species — or what each species uses them for. Some use them for walking, while others use them for swimming or burrowing.

Avoid touching them as these bristles can penetrate your skin easily and cause irritation.

This worm has pushed its head into a tube as it hunts for food. It is about 20 cm long and 1 cm wide.

Coastal Horseshoe Crab

Scientific name: *Tachypleus gigas*

Let's Identify!

I have a peculiar dome-shaped shell
Which looks so unique.
It's known as a **carapace***
This gives me my horseshoe physique.

My carapace has a larger front portion,
And a smaller spiney back section,
With a flexible joint in between.
This carapace is my protection.

I sport a strong stiff tail
That looks like a fierce spine!
However, it's not used for fighting
But to right myself when I am supine.

I have eyes on my front shell and
My tail has a triangular cross section.
It is thicker nearer my body
And the ridge on top has small serrations.

*Let's Pronounce
Carapace:
"Care-Rer-Pace"

Feeding Time!
I move about actively along the coast
And feed on about anything I can grab.
This includes worms, clams, carcasses,
Algae and sometimes, a crab.

Living Fossils
My form and shape have not changed much
Since many millions of years ago.
I'm more related to spiders and scorpions
Rather than true crabs, just so you know.

Fish Bites

There are 2 species of Horseshoe Crabs in Singapore. The other is the Mangrove Horseshoe Crab (Scientific name: *Carcinoscorpius rotundicauda*).

....

The Coastal Horseshoe Crab has 8 other "eyes" on the carapace that are actually sensory organs that detect light.

Serrations on the top of the tail spine

The Coastal Horseshoe Crab's tail spine is as long as its body. It can grow up to 25 cm in length, excluding the tail spine.

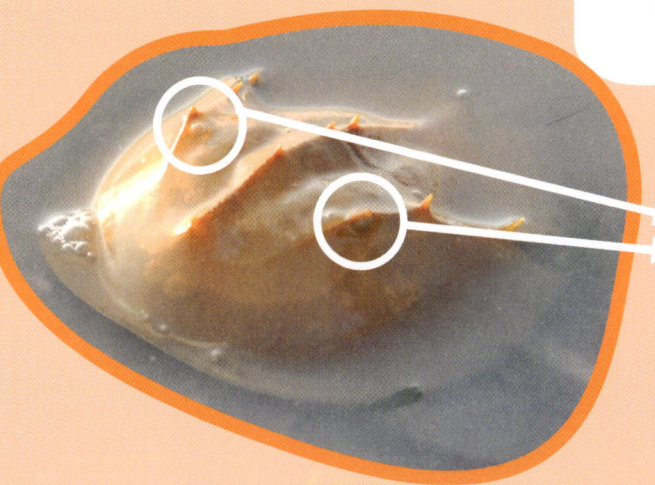

The Coastal Horseshoe Crab has 2 main compound eyes on the top of its carapace for seeing.

31

Orange-Striped Hermit Crab

Scientific name:
Clibanarius infraspinatus

Let's play hide & seek with our Orange-Striped Hermit Crab friends. Scan the QR code to watch the fun video.

Let's Identify!

I am common on the seashore
But I'm always hiding in a shell.
I carry this shell around with me
For it is in this that I dwell.

My legs are slender and elongated,
Sometimes you can see them out.
With brown and orange stripes,
That's how I move about.

2 different *Melongena* shells, both with an Orange-Striped Hermit Crab cautiously coming out. The legs characteristically have brown and orange stripes.

This one has occupied an empty Noble Volute shell.

When you see a shell, watch carefully. The legs under the shell tell you that it is occupied by a hermit crab.

Fish Bites

House Hunters

Hermit crabs have body parts that are soft. This is why they occupy empty shells, which give them needed protection. As they grow in size, they need to look for larger empty shells. When they find one that is suitable, they will discard the present shell and quickly switch to the new one.

Hermit Crabs are not fussy about which type of shell they use. It is the size of the empty shell that matters.

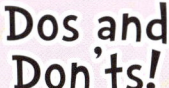

Dos and Don'ts!

Never pull a hermit crab out of its shell.

1. Doing so may cause injury to it.
2. It will be vulnerable running about looking for another shell which has to be empty.
3. With its soft body exposed it will be attacked by predators.

Do not take a hermit crab home because we are unable to provide it with the conditions and food it is used to. All we will do is shorten its life.

33

Ball Moon Snail

Scientific name: *Neverita didyma*

Let's Identify!

My spherical shell is white to off-white,
I can grow 3–4 centimetres in size.
My body can extend many times larger than my shell.
Now, isn't that a magical surprise?

When I fully retract into my shell,
A brownish **operculum***
acts like a door.
Although my shell is polished and shiny,
Don't bring me home
when you see me on shore.

*Let's Pronounce
Operculum: "Oar-Per-Kew-Lerm"
Radula: "Rare-Dew-Lah"

Feeding Time!

I move across or burrow through sand.
My prey are clams and snails.
Do you wonder how I eat them?
Read on for the details!

Many tiny teeth on my tongue called **radula***
Help me drill a small hole through their shell.
Then I release digestive juices to ingest them,
I am a ferocious hunter, as you can tell!

Here's a snail lying half-buried in the sand.

Here's a snail with its foot extended.

Fish Bites

What is an operculum? 'Operculum' is a Latin word which means lid or cover in English. Talk about a literal name!

....

The operculum is a hard structure attached to a snail's foot. When the snail withdraws its foot into the shell, the last part in is the operculum which fits perfectly to cover the shell opening to give it good protection.

Pearl Conch

Scientific name: *Strombus turturella*

Let's Identify!

I can grow to 7 cm in length,
My name, "Conch", is pronounced as "conk".
I am usually brown in colour, and
The locals call me the "gong-gong".

The underside of my shell is white,
And the opening is thick and flared.
My eyes are on the tips of eyestalks,
They are large — but don't be scared.

Feeding Time!

I have a long flexible proboscis
That helps me sense seaweed.
Unlike volutes, conchs are herbivores,
So sea plants make up my daily feed.

This is what a Pearl Conch looks like from the bottom. Look at the flared lip with a thick free edge.

The eyes of a pearl conch are at the tip of the eyestalks.

Fish Bites

Pearl Conchs have an **operculum**. Do you remember what an operculum is?
Hint: It was explained on the Ball Moon Snail page.

· ·

Pearl conchs move by pulling themselves over the sand. The way they do it makes it look like they are pole-vaulting, so it is easy to spot them when they hop.

When upside down, the blade-like operculum is extended to stick into the sand and help the conch to right itself.

This is the front end of the snail.

Here's a top view of the Pearl Conch.

Noble Volute

Scientific name: *Cymbiola nobilis*

Let's Identify!

I have a thick cone-shaped shell,
That is beige, orange, yellow or black.
Patterned with lengths of lines
that look like red or brown zigzag.

Feeding Time!

I move around more actively at night,
Crawling over sand in search of food.
I am a **carnivore**, and clams and snails
Get me in a very hungry mood!

My long tube-like **proboscis**
Sticks out above the sand.
I use it to seek out and catch my prey
Then it's game over for them!

Here's one that is not completely underwater. See how, at low tide, it continues to hunt prey with its foot and proboscis fully extended.

I suffocate my prey with my foot
And this forces it to open its shell.
I then stick my proboscis into its flesh
And slowly scrape my prey out well.

Fish Bites

This species is found only in Singapore and the southern part of West Malaysia.

This is a Noble Volute shell. See its distinct red and brown zigzag lines. This one is empty.

Proboscis

Dos and Don'ts!

Although volutes are harmless, it is best to leave them alone as they may be feeding or looking for prey.

The snail will die if it is forcibly pulled out of its shell.

Here's another that is occupied. Look how it glides over the sand looking for prey when underwater.

Starry-Mouthed Nudibranch

Scientific name: *Bornella stellifer*

Let's Identify!

I belong to the same group as snails
But I come without a protective shell.
My soft body is completely exposed
But I have ways to protect myself well.

I have finger-like projections down my back,
Each one has fine gills that look feathery,
My body is narrow and elongated, and
Brightly coloured orange — making me look fiery!

Acting as my sense of smell,
I have a pair of orange sensory organs.
These are called **rhinophores**
Located on my front-most pair of projections.

The brightly-coloured Starry-Mouthed Nudibranch can grow up to 3 cm long.

Top view of the front shows the star-shaped oral tentacles.

Rhinophore

Oral tentacle

41

Acting as my sense of touch,
I have a pair of star-shaped tentacles
Located on either side of my mouth,
Making me a starry-mouthed spectacle!

I'm Distasteful!

I have **glands** that secrete chemicals
Which do not taste very nice.
A predator that swallows me will spit me out
And next time it sees me, it will think twice!

Feeding Time!

We feed mainly on small **hydroids**
Which are in the same group as corals.
This makes us carnivores
Since hydroids are animals.

The nudibranch's rhinophores are on the first pair of projections. Its oral tentacles on either side of the mouth form a star shape.

Fish Bites

Why bright colours? Although the Starry-Mouthed Nudibranch's bright colours make it easy to detect, being bright also serves as a warning to predators that they are not nice to eat.

Predators that have experienced the distasteful chemical secreted by the nudibranch's glands will learn never to go after it in future.

Eight-Armed Seastar

Scientific name: *Luidia maculata*

Let's Identify!

I usually have **eight** arms,
Sometimes six, seven or nine.
When my arm gets bitten off,
Don't worry, I am still fine.

I have great powers of **regeneration**
For I can regrow an arm.
I am larger than most sea stars
And I think that is my charm.

Let's watch the Eight-Armed Seastar in action! Scan the QR code on this page to watch the video.

45

Feeding Time!

My mouth is on my **underside**,
And I feed actively at night.
Small crabs, snails, clam and worms
I swallow them whole at first sight!

Tube Feet

Movement Made Easy

Tube feet under me help me move about;
Small spines on my sides give me grip.
Paxillae on top give protection and flexibility
And these are all I need for a trip!

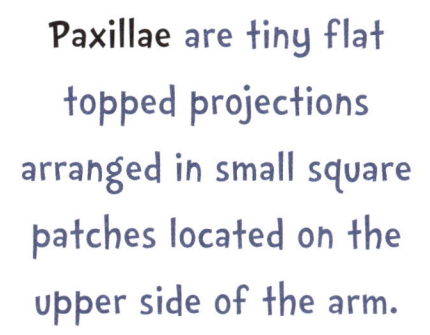

Paxillae

Fish Bites

Paxillae are tiny flat topped projections arranged in small square patches located on the upper side of the arm.

In case the seastar gets overturned by strong waves, its tube feet, spines and paxillae all help it to quickly right itself!

Biscuit Sea Star

Scientific name:
Goniodiscaster scaber

Let's Identify!

I have short arms with blunt rounded tips.
Squat and compact, I'm as cute as a button.
I have patterns in shades of brown and orange.
I look like a cookie fresh from an oven!

My underside is pale and darkens towards the mouth,
With bluish grooves under each arm.
Orange tube feet protrude from the grooves
But I don't cause humans any harm.

This one shows a different colour pattern.

Feeding Time!

But clams, snails, sponges and corals beware!
Sea squirts too, for I love to feed on these.
I use my tube feet to attach myself
And won't let them go even if they say "please".

I **evert** my stomach through my mouth,
Cover my prey and start digesting.
Stomach enzymes help me digest them
And when done, I pull my stomach back in.

The Biscuit Sea Star has short arms with blunt tips. There are no spines along the sides of the arms.

Fish Bites

What does it mean to evert something? "Evert" is a word used in biology to mean "inside out". Everting the stomach means to push the stomach out of the body through the mouth so that digestion takes place externally.

....

A Biscuit Sea Star's size can range from 2–15 cm in diameter.

The underside is pale with a darker centre. Grooves under each arm are bluish with numerous orange tube feet.

Painted Sand Star

Scientific name:
Astropecten vappa

Different Painted Sand Stars have different patterns on their upper side. Their patterns are all very striking.

Let's Identify!

I can grow to 10 cm in diameter.
I'm quite beautiful, I think.
I come in a bluish hue
And I look quite distinct.

I have striking patterns on my upper side
With dark blue or orange lines.
My arms appear extra broad because
At each side, I have long white spines.

Feeding Time:

I hunt for small snails and clams
And I feed actively at night.
I move fast and catch them
Before they can take flight.

I do not **evert** my stomach.
Unlike my Biscuit Sea Star friends,
I swallow my prey whole, then
Spit out the remains when digestion ends.

Fish Bites

Tube feet on the Painted Sand Star's underside help it move actively over the sand.

....

It can also dig itself into the sand fairly quick.

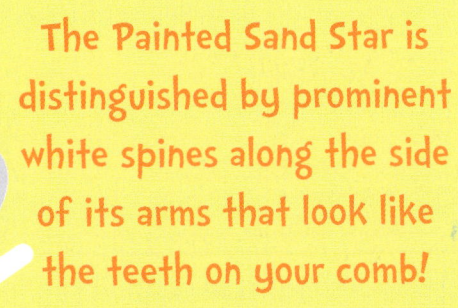

The Painted Sand Star is distinguished by prominent white spines along the side of its arms that look like the teeth on your comb!

Here's a Painted Sand Star in the process of digging itself into the sand. Doesn't it look like someone painted a star in the sand?

51

Knobbly Sea Star

Scientific name: *Protoreaster nodosus*

Let's Identify!

I'm attractive and brightly coloured.
Ranging from red, to orange, beige and bluish-brown,
I can grow up to 30 cm in diameter —
I'm one of the largest sea stars here to be found.

Cone shaped nodules on my upper surface
Have earned me my name as Knobbly.
Now that you know what to look out for,
You'll know me when you see me, probably.

My nodules have many different functions.
First, they help scientists with identification.
Next, they provide me with protection.
Finally, they also help me with respiration.

The Knobbly Sea Star is recognisable due to the cone shaped nodules on its upper surface.

Feeding Time!

My mouth is on my underside
And I eat many things in the sea.
Microbes, snails, corals, sponges
And **detritus** — also known as debris.

The pattern of nodules on individual Knobbly Sea Stars vary. Researchers use these specific patterns to recognise individuals.

They usually have 5 arms. The one on the right has lost an arm, but it will regrow with time.

Fish Bites

Do you remember what "evert" means?

This term was explained on the Biscuit Sea Star page. The Knobbly Sea Star can also evert its stomach and envelop and digest its prey externally. This way, it can prey on large animals.

....

Predators of the Knobbly Sea Star include Pufferfish and Triggerfish. These stars are also preyed upon by the very large Triton snail, which is not present in Singapore.

Cake Sand Dollar

Scientific name: *Arachnoides placenta*

Here's the underside of a Cake Sand Dollar. Do you see its mouth at the centre?

Let's Identify!

You may find me lying flat on the beach
So be careful, as I can be quite brittle.
I can also be found in a vertical position
With part of my body sticking out a little.

I resemble a stiff, round, flattened disc
With my mouth at the centre on my **underside**.
My body is covered with small, fine spines
Which help me move around and hide.

I use my spines to dig edgewise into the sand
So my spines help me with **locomotion**.
Spines on my underside move food to my mouth while
Spines on my upper side help me with **respiration**.

Feeding Time!

I like to feed on algae, plankton
And tiny remains of dead sea creatures.
I am prey to snails, sea stars and carpet anemones
Oh, shorebirds too, when they find me on the beaches.

This is the upper side of a Cake Sand Dollar. It can grow to about 7 cm in diameter.

Here's a picture of the test of a dead Cake Sand Dollar in the upside-down position.

Fish Bites

When alive, the Sand Dollar's body is covered with small fine, flexible spines.

When dead, the Sand Dollar's spines drop off, leaving a skeleton known as a "test" that is the same size and shape of the body.

Thorny Sea Cucumber

Scientific name: *Colochirus quadrangularis*

Let's Identify!

My body is coloured from orange to red,
With long blue, green or grey bands
I am small and tube-shaped,
With tube feet to crawl over the sands.

The blunt horn-like projections on my body
Are soft and harmless to touch.
They're why I'm named "Thorny",
But I really don't hurt people much.

Feeding Time!

My mouth is found on my front end.
And when I want to have a snack,
A ring of yellow and red tentacles extend,
Trap plankton, then into my mouth they retract.

The Thorny Sea Cucumber is called such because of its many horn-like projections.

Fish Bites

On the back end of the sea cucumber is its anus, which is surrounded by 5 short projections shaped like a star!

••••••••••••••••••••

The Thorny Sea Cucumber is usually between 5 and 10 cm in length.

This Thorny Sea Cucumber is waiting for the tide to rise before extending its tentacles to filter feed.

Here's a Thorny Sea Cucumber extending its bright yellow and red feeding tentacles from its mouth to feed. It has five short projections surrounding its anus.

Anus

Mouth

Pink Warty Sea Cucumber

Scientific name: *Cercodemas anceps*

Hi! I'm a Warty!

Let's Identify!

I am small and tube-shaped,
With little blunt warts on my body,
My colours are pink and yellow
That's why I'm called "Pink Warty".

I can be found in large groups
That may look like a field of pink in your eyes.
I can grow to about 8 cm in length
But I can also be very small in size.

Dos and Don'ts!

The Pink Warty Sea Cucumber looks so colourful and cute! Is it safe to touch it?

Although the Pink Warty is harmless, it is best to avoid handling it.

Like most sea cucumbers, they secrete chemicals such as saponins for their protection. If you have to touch them, be sure to thoroughly wash your hands as soon as you can.

Ever wondered how the Pink Warty Sea Cucumber moves? Scan the QR code on this page to learn more!

Feeding Time!

Feeding tentacles surround my mouth.
When I am in water, they extend
To trap plankton and small creatures,
Then I pull them into my mouth like a hand!

When my tentacles extend,
I look like I am wearing a crown.
They're commonly yellow with red spots.
I'm usually the talk of the town.

A field of pink. Pink Warty Sea Cucumbers can be found in large groups like this one.

Pink Warty Sea Cucumbers with their tentacles extended, creating a feathery ring surrounding the mouth.

A case of mistaken identity?!

Which one is the Pink Warty Cucumber?

Fish Bites

Both the Pink Warty and Thorny Sea Cucumber can be found together and are easily mistaken for each other, but look closely and you can tell them apart.

•••••••••••••••••••••••••

The Pink Warty Sea Cucumber has pink warty bumps instead of blunt conical 'horns'. It also has the colour yellow on its body, unlike the Thorny Sea Cucumber.

•••••••••••••••••••••••••

Movement Made Easy

On the underside of the Pink Warty Sea Cucumber, there are many small pink tube feet. They are arranged in 3 rows from front to end. This enables the Pink Warty to move about easily.

Smooth Sea Cucumber

Scientific name: *Acaudina* sp.

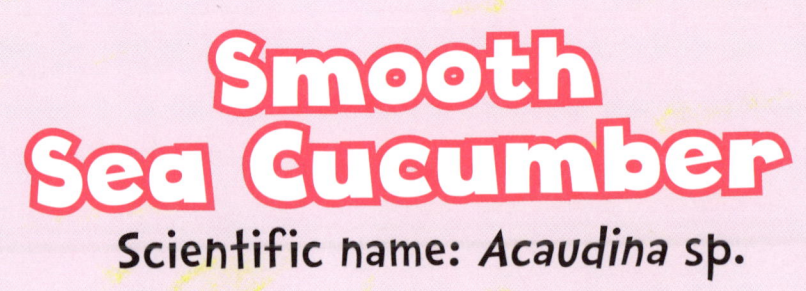

Let's Identify!

Did you know that I am quite common?

But I might be easily missed

Because I like to lie buried in sand

But I actually do exist!

Scan this QR code to head to Changi Beach and see the Smooth Sea Cucumber in action!

My body looks like a fat tube
And my mouth is at one end.
I'm purplish, sometimes orange with spots.
But I'm usually covered with muddy sand.

I'm soft and slimy to the touch
And I look smooth and shiny.
I have no bumps or warts
And I'm definitely not spiny.

Feeding Time!

I will push myself out of the sand
If I need to move about.
Hey, this cucumber needs to eat,
So I'll definitely come out to scout.

Chubby feeding tentacles surround my mouth.
Small animals, algae and plankton are to my taste.
I move around in search of them but
I'm also a scavenger —
I won't pass on eating waste!

Fish Bites

Smooth Sea Cucumbers are quite large and can range from 10–15 cm in length.

．．．．．．．．．．．．．．．．．．．．．．．．．．．

The mouth is circular and gives the cucumber an "oooooo" look. The short chubby tentacles that surround the mouth appear to pulsate continuously.

White Salmacis Urchin

Scientific name:
Salmacis sphaeroides

Let's Identify!

I measure about 8 cm in diameter,
I look like a dome sitting on the sands.
I have short, stiff white spines
Sometimes marked with maroon or green bands.

My skeleton, also known as a test,
Protects my soft parts that animals can eat.
On the test protrude white spines
And many slender flexible tube feet.

Movement Made Easy

The spines and tube feet on my upper side
Constantly wave about very actively.
But I also have tube feet on my underside,
So I can crawl over sand quite rapidly.

The White Salmacis Urchin has short, stiff, sharp spines all over its body. The spines of this one has maroon bands.

The tube feet and spines on the body help to move and position materials on the urchin.

The White Salmacis Urchin sticks materials to the upper side of its body. Materials include seaweed, dead shells and small rocks. These are attached to the body to help camouflage it or prevent it from being tipped over by strong waves.

Fish Bites

Seashore Fashion or Smart Camouflage?

This sea urchin is a collector of anything it comes across — e.g., dead shells, pebbles, leaves and seaweed. It will stick these items onto the upper side of its body. But it does not do this just for fun or fashion. It is preyed on by fish and large snails. Decorating its body with items from its surrounding helps to camouflage itself from predators.

Acorn Worm

There are more than 100 species worldwide. The species inhabiting our shores has yet to be identified. This is why conservation is important as it allows scientists time to study the biodiversity that Singapore has.

Let's Identify!

I'm called a worm and I may look like one,
But I'm really not considered a worm,
Because my body is not segmented.
Still, my looks might make you squirm.

I'm related to the **Echinoderms***
Which include the urchin and the sea star.
I am what you call a **Hemichordate***,
Wow! Doesn't this fact sound bizarre?

I live all over the beach —
But "Where are you?" you might ask.
I burrow deep so I'm hardly seen
But I leave coils of my worm cast.

A spiral coil of worm cast produced by the Acorn Worm when the tide is out. What is "worm cast" you ask? Find out in this entry's Fish Bites!

Dos and Don'ts!

Do not attempt to dig up the worm cast mounds as the burrow is deep and the chances of killing the Acorn Worm in the process are high.

Here's a cast beginning to disintegrate as the tide comes in.

*Let's Pronounce Echinoderms:
"Air-Kai-No-Derms"

Hemichordate:
"Hair-Mee-Kor-Date"

Eco-Ambassador!

Look for mounds of sand arranged in coils
For these are made by me, my friend.
I consume mud and sand for nutrients and
What I don't need is expelled out my other end.

My actions play an ecological role,
As the layers of mud and sand that I displace
Get enriched and oxygenated
When they are moved to the surface.

Some mounds of worm cast are really large and tall!

If you look really closely, the orange hind end of the Acorn Worm can be seen here, as it expels its worm cast. This is about all you may ever get to see of the animal!

Fish Bites

What is "worm cast"?

As the Acorn Worm burrows, it consumes sand and mud to absorb organic material. It then expels the sand and mud through its hind end. This expelled material is called the "cast", otherwise referred to as the worm's waste materials.

A short cone-shaped proboscis that looks like an acorn is at the front of the body, which explains the Acorn Worm's name. It lives in a U-shaped burrow and some can grow to as long as 2 metres.

Let's Discover, Singapore! (Set)

Hello little readers! Did you know that there's something for you to discover in every little corner of Singapore? Let's Discover, Singapore! Set is a book set designed to encourage young readers in Singapore to observe and learn from the world around them. Comprising two titles, Let's Discover Science, Singapore! and Let's Discover Our Seashores, Singapore!, the books take junior explorers all across the city state—from the shiny man-made structures like the Esplanade, to the architectural wonders of nature right at the doorstep of our little island. The reader experience for both titles is enhanced with bonus features made possible through augmented reality or QR codes.

To receive updates about children's titles from WS Education, go to https://www.worldscientific.com/page/newsletter/subscribe, choose "Education", click on "Children's Books" and key in your email address.

Follow us @worldscientificedu on Instagram and @World Scientific Education on YouTube for our latest releases, videos and promotions.